The path of the electron

ELECTRONICS

MARCOS CERVANTES JANSSEN

ELECTRONICS

THE PATH OF THE ELECTRON

By: Marcos Cervantes Janssen

INDEX:

- FOREWORD: 6
- RESISTANCE: 8
- INDUCTOR OR COIL: 11
- CAPACITOR OR CONDENSER: 14
- DIODE: 17
- TRANSISTOR: 19
- INTEGRATED CIRCUIT: 22
- EPILOGUE: 23

First edition: August 7 , 2022

Copyright © *2022 Marcos Cervantes Janssen*

Edited by Editorial letr@Roja

https://newtekjanssen.es.tl/letra3roja@gmail.com

https://www.facebook.com/LETRA3ROJA

https://www.newtek.janssen@gmail.com

https://payhip. com/letra33roja

PROLOGUE:

Understand in a really practical way, the components, symbols, concepts and essential parts of this wonderful science, which is present in all areas of our lives, education, medicine, cooking, entertainment, home, business, communications, and many more known, which in daily life make a great contribution to our efficiency in different areas. The path of the electron through different materials, results in many useful applications for humanity, the electron being an existing electrical particle and that in a set of thousands, produces a current, called electricity, this current is transformed and produces different phenomena when manipulated by the components in the different modules that we will see here.

Electrons are part of atoms, particles that make up everything that exists in the universe, in such a way that each electronic component responds differently to the passage of electrons or electric current, this energy flows through a potential difference. I will present you with a clear and direct language, the events that occurred in each component and their practical applications, in the daily use of life, so you will understand, and enter, in a very interesting world, current and future without a doubt. Today all communication and the computer world depends directly on these precise movements of the electron through computer circuits and other devices that have circuits full of components, presence, absence, and interaction, is the subject that concerns us in this treatise, we will see basic and simple formulas for a better understanding of this topic that is so necessary and current.

RESISTANCE:

The resistance is a component, which as its name indicates resists, in the case of the electron and its passage through it, the resistance stops the flow of them, it is in a practical way, like crushing a hose, or closing partially the faucet, resulting in a decrease in the flow of water, it is in this way that the electrical flow is reduced, this means that fewer electrons pass through a resistance, as this is of higher resistance value, the Reducing the electrical flow allows us to control the diverse effects of the electrical current in our tools and obtain the results desired by our design, thus taking advantage of the changes in temperature and electrical flow.

Controlling the voltage is of vital importance for the digital subject because we know that the logical 1 in binary mathematics is a voltage of 5 volts and its absence is the binary zero, with which this wonderful world of bits is treated, by the definition of the present voltage.

The resistance when exerting said action to the passage of electrons, produces heat, a thermal effect that represents the transformation of energy, a practical example are the basic heaters to treat water in a bucket, let us also remember the rear window defogger system in automobiles , and as another common example the clothes iron.

Each resistor according to its size and value produces this thermal effect according to the value and electrical flow through it.

If the admitted current for each type of resistance is exceeded, it melts like a fuse, that is why the work specifications must be known. The resistance in electronic circuits is of a millimeter dimension, and in industry, of considerable sizes, made up of a filament, which inside a ceramic body, has the chemical properties of its constitution that oppose the electric current. The electric current, not being able to pass linearly as its nature proposes, is transformed into heat, thus the body of the component is affected by its external temperature.

In a series circuit, the resistances add up, but in a parallel connection, we must use a special calculation formula. There is a color code which reveals in the body of the electronic resistors, its functional value, all these data are necessary for the design of modules in electronics.

INDUCTOR OR COIL:

This component is an element, very similar to the resistance, but its operation is not to oppose in this case the electric flow, but to oppose its variations, being very important for communications and signal filtering. frequency.

We will see in the components two types of electric current, the polarized one called direct current, and the alternating one that flows with a frequency of change of poles.

The coil is a spiral that when conducting direct current, does not exert opposition and behaves as a linear conductor, not so for alternating current, in this case the coil opposes the alternating current, and a magnetic field is created around the component, thus the voltage is converted to magnetism instead of heat as was the case with resistance.

An example of the use of coils is radio tuning, in which it is vitally important to convert electrical signals into magnetic ones, and in this way, be able to transmit them aerially, in the same way, the receiver with the appropriate formulas, can tune these signals. aerial created, and receive the information contained, at considerable distances.

Having a transmitter, and the opportunity, that it can be received, through many modules at the same time, was the great boom of the radio, as well as the tuned communications, provided a correct dialogue between two specific points.

Inductors today are used in very powerful applications such as microwaves and aerial voltage transmission, such as the new magnetic induction stoves, which offer low consumption and high thermal efficiency.

So we have the tesla antenna, which is a coil with a special configuration, and a very interesting construction. It converts the electric current into a dense magnetic field, called inductive plasma, which is not only a signal carrier, but also an air voltage emitter currently used for wireless electronic charging of different devices, the applications of inductors will continue to develop. in the metallurgical, space and medical industry, since its importance when inducing is fundamental in high voltage wireless projects, together with computer transmission, plus what is revealed over time, are today the subject of study, and discovery of innovations truly necessary for shaping our future in wireless power, this for quantum space exploration.

CAPACITOR OR CONDENSER:

This component, unlike the coil, opposes the direct current, and not the alternating one, it is called a filter for its function of stabilizing the voltage when it has variations, it basically works as a high speed accumulator of loading, and unloading.

There are two types of capacitor, the ceramic for high alternating frequencies, and the electrolytic, called a filter, which is also polarized at its terminals, which is used to suppress voltage spikes, called noise.

In digital electronics, it is of vital importance that the signal is very stable, since all the data is based on zero volts, and 5 volts, so that a voltage spike will escape without having been assigned, if not only due to technical failure. , causes a false binary encoding and decoding, when unreadable or increased bits are encountered.

Condensers in audio systems provide professional definition, as well as the necessary equalization with the help of coils and resistors, for a specialized sound output, thereby achieving all the specific levels and ranges necessary for each instrument and vocal tone. , so the capacitor is vital.

The capacitor in combination with other components, generate a signal called square, which is a compass that defines the consecutive march, in digital processes, thus the programming is completely based on synchronized data packets, it is therefore for this matter that the definition of a repetitive signal, but of great quality in its lapses, will lead to the information being generated, transmitted and read efficiently, in each and every one of the digital equipment that we know.

There is a variable capacitor, which within a circuit, makes a series of changes that allow tuning different frequencies, or generate them as the case may be, so in digital systems digital and analog information can be generated, in very precise ways to achieve speed and safety in this area.

Also within the subject of memory, it is really necessary that the capacitors retain their precise levels at the exact times, to contain the precise information in the memory devices and at effective speeds.

On the subject of electronic coupling, capacitors perform very important tasks, there is a piezoelectric capacitor, capable of generating electric current, if physically pressed, with certain frequencies and specific purposes.

DIODE:

The diode is a very useful electronic component, with it the transistor is manufactured, the diode conducts the direct current only in one direction, and the alternating current rectifies it, because by conducting only one pole, we will have only one sign to the left. output of the device, this is called rectification.

The diode works the same as a hydraulic prick, or the so-called check key, it only allows flow in one direction, thus the electronic component is also used in digital electronics to identify zeros or ones.

Rectifier diodes are used in all power supplies on the market, their property of driving only in one direction, correcting and rectifying wrong polarization in precision circuits, each diode represents a typical consumption of 0.7 volt.

There is a special diode with the name zener in honor of the inventor, this diode has the peculiarity of being part of the AM receiver, which does not require batteries, this is possible, because due to its sensitivity, it works with only the working voltage of the diode , which is the signal that enters the circuit above the air through its antenna and this signal is converted into modulated voltage by the diode, so with the help of a piezoelectric earphone, it will be heard clearly enough.

Diodes currently make up integrated circuits, thus combinational logic is possible to work through these components that are the basis of logic gates, thus together with other basic components they form the foundation of miniaturized semiconductors, and compacted by thousands, in these tiny devices, this component is indispensable.

TRANSISTOR:

This component has three terminals, called base, collector and emitter, with the base controlling the component, and the collector and emitter fulfilling the main function of the component.

The main function of this component is to be a switch, regulator and gate in combinational electronics, thus this element of electronics becomes the first integrated circuit developed.

Let us suppose to understand, in a practical way, that the component that is a stopcock in a water pipe, the crank being the base, the water inlet the collector and the emitter outlet, so it is depending on the movement of the crank, that is, the base, which in electronics is controlled by the number of volts supplied to the base, is thus the current between collector and emitter.

Transistors are the technical basis of complex microchips, their switching property is the one who performs the combinational logic necessary for digital development, thus miniaturized, and arranged in logic gates in essence, thus we will have with this the first foundation of computational electronics, of already three decades in development, so it will continue in this way.

We also have power transistors, which are for industrial purposes, so the industry has been automated and with these components such as switches and regulators, which control the processes of different fields in the industry.

Also in the area of sound transistors they have developed powerful amplifiers, they are part of complex equalization and modification systems to improve this activity.

Within the field of measurements and instrumentation, transistors in combination are a series of sensors, they are today great medical and industrial tools in the special area of measurement, also of controlled supplies, in remote systems, for mining, medicine and the space area.

Transistors, whether control or power, are becoming more advanced, through research, which provided further increases in precision, performance, as well as improvements by reducing their size, and increasing their efficiency.

Today the number of transistors encapsulated in microchips is growing exponentially, due to the evolution of their manufacture and improvements in their designs, which in the future will be like neural networks.

INTEGRATED CIRCUIT:

It is here where all the components evolve individually, reducing their size and increasing their efficiency, thus, in a set for special purposes, they are interconnected within an encapsulation, called an integrated circuit.

The so-called IC is the very center of computing, each motherboard has a microchip as a processor unit, with countless elements, made up of very specialized designs and made for high-precision purposes.

For practical reasons the size of our modern technology is getting smaller every day, but more powerful, we have realized that the purpose of each of these complex components is evolving in an incredible way.

EPILOGUE:

We know that our human body is made up of different systems, in which we find the blood circulatory system, and the nervous system, in the latter, there is an electric current circulating through our body.

Our neurons are also charged with energy, which circulates without stopping, throughout our lives. In the wonderful topic of electronics, we realize how we have learned from nature, and reproduced many of the functions that it gives rise to, all this has led us by the hand, to generate computers, it is here that we artificially develop , a programmed artificial mind, which has the objective of learning and making more and more of its own decisions, by accumulating experience. More along with the sector of monitoring and giving life to humanoids with robotic limbs, they are given these, artificial minds, thanks to microchips.

Be they computers, houses and even humanoids, we will have robots in the future, capable of reacting to principles of respect and mutual help. Let us remember that we are its creators and for this reason we are responsible for its learning, evolution and legacy. It is in this way that plants and animals evolve and that the human being, through his creation, can really advance in the total knowledge of creation. It is exciting that each component is totally different, but all of them are essential, and have a real place just for them. The updates of their manufacture take us by the hand to the future, being the electronics of the future, the fact of integrating our biological reality , its proper integration and true human respect. For this reason, we must always recognize that electronics is the study of natural processes recreated by our hands for the benefit of the common good.

All rights reserved. Under the sanctions established

in the legal system,

without written authorization from the holders of *Copyright* ©

the total or partial reproduction of this work by

any means or procedure

, reprography and computer processing

.

Hello, I am a researcher, writer and communications engineer, throughout my life, I have experienced strong situations in every way, I wish that your life goes better and better, and that you evolve as much as you can by expanding your knowledge, mind and will, I am sure we can find an expand our existence, I want to accompany you always, and I thank you in advance YOU ARE

Understand in a really practical way, the components, symbols, concepts and essential parts of this wonderful science, which is present in all areas of our lives, education, medicine, cooking, entertainment, home, business, communications, and many more known, which in daily life make a great contribution to our efficiency in different areas. The path of the electron through different materials, results in many useful applications for humanity, the electron being an existing electrical particle and that in a set of thousands, produces a current, called electricity, this current is transformed and produces different phenomena when manipulated by the components in the different modules that we will see here.

Electrons are part of atoms, particles that make up everything that exists in the universe, in such a way that each electronic component responds differently to the passage of electrons or electric current, this energy flows through a potential difference. I will present you with a clear and direct language, the events that occurred in each component and their practical applications, in the daily use of life, so you will understand, and enter, in a very interesting world, current and future without a doubt. Today all communication and the computer world depends directly on these precise movements of the electron through computer circuits and other devices that have circuits full of components, presence, absence, and interaction, is the subject that concerns us in this treatise, we will see basic and simple formulas for a better understanding of this topic that is so necessary and current.

ISBN 9798362993047

www.ingramcontent.com/pod-product-compliance
Lightning Source LLC
Chambersburg PA
CBHW050327220526
45465CB00005B/2170